W0082804

Technical Evaluation Report

EVALUATION FINDINGS FOR ENIDINE, INC. VISCOUS DAMPER

Prepared by the
Highway Innovative Technology
Evaluation Center (HITEC)

A Service Center of the Civil Engineering
Research Foundation (CERF)
CERF REPORT: HITEC 99-02
#40403
February 1999
Product 21

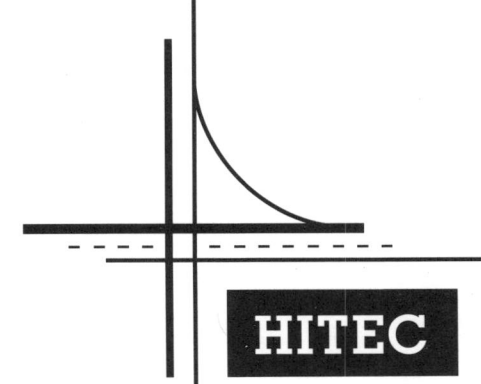

HITEC

Abstract

The Highway Innovative Technology Evaluation Center (HITEC) serves as a clearinghouse for implementing highway innovation by providing nationally-focused, collaborative evaluations of new products and technologies.

This report, *Evaluation Findings for Enidine, Inc. Viscous Damper*, presents the results of a detailed evaluation for one technology out of eleven that were evaluated in this program. The evaluations were designed to test the performance of seismic isolators and dampers produced by several manufacturers.

Library of Congress Cataloging-in-Publication Data

Evaluation findings for Enidine, Inc. viscous damper.
     p. cm. -- (Technical evaluation report) (CERF report; HITEC 99xx)
     "Prepared by the Highway Innovative Technology Evaluation Center (HITEC)."
     "November 1998"
     ISBN 0-7844-0403-8
     1. Bridges--Bearings--Testing. 2. Dynamic testing. I. Highway Innovative Technology Evaluation Center (U.S.)
     II. Series. III. Series: CERF report; 99xx
     TG326.E86242 1998
     624'.252-dc21
                                        98-44498
                                        CIP

The material presented in this publication has been prepared in accordance with generally recognized engineering principles and practices, and is for general information only. This information should not be used without first securing competent advice with respect to its suitability for any general or specific application. The contents of this publication are not intended to be and should not be construed to be a standard of the American Society of Civil Engineers (ASCE), or its research affiliate, the Civil Engineering Research Foundation (CERF) and are not intended for use as a reference in purchase specifications, contracts, regulations, statutes, or any other legal document.

No reference made in this publication to any specific method, product, process, or service constitutes or implies an endorsement, recommendation, or warranty thereof by ASCE or CERF. ASCE and CERF make no representation or warranty of any kind, whether express or implied, concerning the accuracy, completeness, suitability, or utility of any information, apparatus, product, or process discussed in this publication, and assumes no liability therefore. Anyone utilizing this information assumes all liability arising from such use, including but not limited to, infringement of any patent or patents.

Photocopies. Authorization to photocopy material for internal or personal use under circumstances not falling within the fair use provisions of the Copyright Act is granted by ASCE to libraries and other users registered with the Copyright Clearance Center (CCC) Transactional Reporting Service, provided that the base fee of $4.00 per article plus $.50 per page is paid directly to CCC, 222 Rosewood Drive, Danvers, MA 01923. The identification for ASCE Books is 0-7844-0403-8/98. $4.00 + $.50 per page. Requests for special permission or bulk copying should be addressed to Permissions & Copyright Department, ASCE.

Copyright © 1998 by the American Society of Civil Engineers.
All Rights Reserved.
Library of Congress Catalog Card No: 98-44498
ISBN 0-7844-0403-8
Manufactured in the United States of America.

Acknowledgments

The Highway Innovative Technology Evaluation Center (HITEC), a service center of the Civil Engineering Research Foundation (CERF), prepared this report. HITEC wishes to acknowledge the special contributions of individuals whose efforts and suggestions have significantly influenced this report. Notably, this report is based on work by members of a technical evaluation panel who volunteered to develop an evaluation plan and carry out its objectives. The panel is composed of Chairman Mohsen Sultan, P.E., California Department of Transportation (Caltrans); Steve Bradford, P.E., Alaska Department of Transportation; Maria Feng, Ph.D., University of California, Irvine; Hamid Ghasemi, Ph.D., Federal Highway Administration; Stewart Gloyd, P.E., Parsons Brinckerhoff, Quade & Douglas; Roy Imbsen, Ph.D., P.E., Imbsen & Associates; Salah Khayyat, P.E., Illinois Department of Transportation; Myint Lwin, P.E., S.E., Washington State Department of Transportation; Ayaz Malik, P.E., New York State Department of Transportation; Rolando B. Nimis, P.E., Federal Highway Administration; Walter Podolny, Jr., Ph.D., P.E., Federal Highway Administration; Charles Seim, P.E., T.Y. Lin International; Li-Hong Sheng, P.E., Caltrans; Arun Shirole, P.E. (former panel member), formerly with New York State Department of Transportation; and Edward Wasserman, P.E., Tennessee Department of Transportation. The efforts of Armand Onesto at the Energy Technology Engineering Center (ETEC) should also be noted, as he was responsible for overseeing the testing of the seismic isolators and dampers.

The writing efforts of Dr. Ghasemi, Mr. Sheng, and Mr. Onesto should also be noted as they were responsible for preparing the text of all the reports developed for this project.

Caltrans was instrumental in the completion of this project as they oversaw the testing program and took the lead in analyzing the test results.

Among the staff who worked on this project, I wish to acknowledge the special efforts of HITEC's Director and CERF Vice President, J. Peter Kissinger; Kathleen Almand, P.E.; Michael S. Higgins, P.E.; and Stacy Warner, who were all instrumental to the completion of this very important project.

Publication of this report is made possible, in part, through the contributions by members of CERF's New Century Partnership:

- Black & Veatch
- CH2M Hill Ltd.
- Charles Pankow Builders
- Charles J. Pankow Matching Grant
- Kenneth A. Roe Memorial Program
- Lester B. Knight & Associates, Inc.
- Parsons Brinckerhoff, Inc.
- The Turner Corporation

Harvey M. Bernstein

Harvey M. Bernstein
President
Civil Engineering Research Foundation (CERF)

Disciamer

This document is based on work supported by the Federal Highway Administration under Cooperative Agreement No. DTFH61-93-X-00011.

Any opinions, findings, and conclusions or recommendations expressed in this publication are those of the Highway Innovative Technology Evaluation Center (HITEC) and do not necessarily reflect the view of the Federal Highway Administration.

This report is the result of an impartial, consensus-based approach to evaluating innovative highway technology in accordance with the HITEC Technical Protocol. The data presented are believed accurate and the analyses credible. The statements made and conclusions drawn regarding the product evaluated do not, however, amount to an endorsement or approval of the product in general or for any particular application.

Technical Evaluation Panel Key Contacts

Product: Energy Dissipator

Chair: **Mohsen Sultan, P.E.**
Senior Bridge Engineer
Office of Earthquake Engineering
Engineering Service Center
California Department of Transportation

Panelists: **Steve Bradford, P.E.**
Chief Bridge Engineer
Alaska Department of Transportation

Maria Feng, Ph.D.
Professor
University of California, Irvine

Hamid Ghasemi, Ph.D.
Research Structural Engineer
Federal Highway Administration

Stewart Gloyd, P.E.
Senior Engineering Manager
Parsons Brinckerhoff, Quade & Douglas

Roy Imbsen, Ph.D., P.E.
President
Imbsen & Associates

Salah Khayyat, P.E.
Chief, Bridge Standards and
Specifications Unit
Illinois Department of Transportation

Myint Lwin, P.E., S.E.
Bridge and Structures Engineer
Washington State Department of
Transportation

Ayaz Malik, P.E.
Associate Civil Engineer
New York State Department of
Transportation

Rolando B. Nimis, P.E.
Regional Structural Engineer
Federal Highway Administration

Walt Podolny, Jr., Ph.D., P.E.
Senior Structural Engineer
Federal Highway Administration

Charles Seim, P.E.
Senior Principal and
Senior Bridge Engineer
T.Y. Lin International

Li-Hong Sheng, P.E.
Senior Bridge Engineer
Office of Earthquake Engineering
California Department of Transportation

Arun Shirole, P.E. (former panelist)
Former Deputy Chief Engineer, Structures
New York State Department of
Transportation

Edward Wasserman, P.E.
Civil Engineering Director, Structures
Division
Tennessee Department of Transportation

Client: **Enidine, Inc.**
184 Technology Drive
Suite 201
Irvine, California 92618
Phone: 949-727-9112
Fax: 949-727-9107

Thomas A. Zemanek
Seismic Program Manager

HITEC Project Managers: **J. Peter Kissinger**
Kathleen Almand, P.E.
Michael S. Higgins, P.E.

Consultants: **Energy Technology Engineering Center**
Larry Lowe, P.E.

Contents

Figures

Tables

Preface

The introduction of new or innovative technology to the highway community usually requires that a new product be demonstrated to many, if not all, state highway agencies. This practice is inefficient, time consuming, and often costly, particularly for small companies and entrepreneurs. To overcome these barriers, the Highway Innovative Technology Evaluation Center (HITEC) was established in 1994 in cooperation with the Federal Highway Administration (FHWA), the American Association of State Highway and Transportation Officials (AASHTO), and the Transportation Research Board (TRB). HITEC's mission is to accelerate the process of introducing technological advances to the highway community.

HITEC facilitates the conduct of consensus-based, nationally accepted performance evaluations of new or innovative technologies for the highway community. While the term "new or innovative technologies" connotes "high tech" products often associated with the computer industry, HITEC is available to evaluate almost any product, system, service, material, equipment, or other technology that the owner believes can be used on the nation's highways.

In the case of the seismic evaluation process, HITEC used a specially modified version of the HITEC process designed for group evaluations, which is illustrated below in Figure P.1.

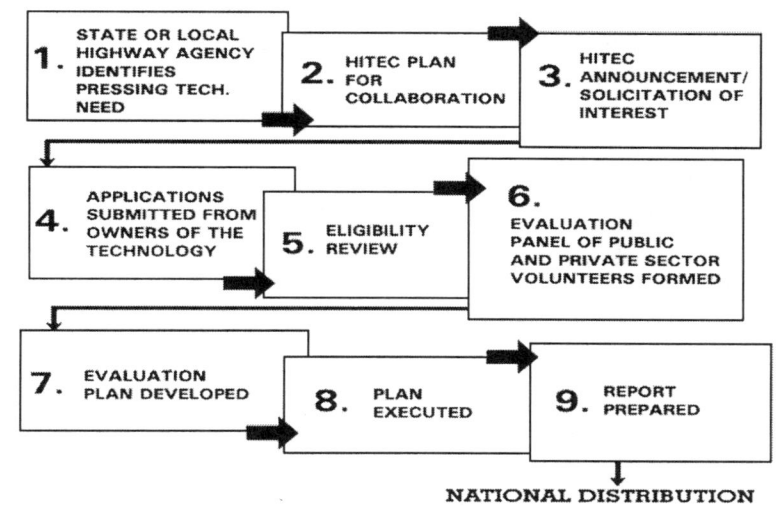

Figure P.1: HITEC Group Evaluation Process

For this evaluation, HITEC collaborated with the California Department of Transportation (Caltrans) and FHWA in the development of the Evaluation Plan. Subsequently, once the Evaluation Plan was developed (HITEC 96-02 [#40162], *Guidelines for the Testing of Seismic Isolation and Energy Dissipating Devices*), letters were sent inviting all known manufacturers of seismic isolation and energy dissipation devices worldwide to participate in the evaluation program.

During the process of soliciting applicants to the program, HITEC staff assembled a Technical Evaluation Panel composed of representatives from the user community, academia, and the private sector. The Technical Evaluation Panel, with the cooperation and assistance of the applicants, identified the specific issues and concerns requiring resolution for these products to be adopted by the highway community. The Technical Evaluation Panel oversaw the development and execution of the Evaluation Plan.

All of the testing to implement the Plan was conducted by the Energy Technology Engineering Center (ETEC), a research testing center managed by the Rocketdyne Division of Boeing Corporation (formerly Rockwell International), which has over 30 years of experience in testing large scale components, including isolator and energy dissipating systems.

Caltrans' engineering staff managed the evaluation program and took the lead in analyzing the test results and preparing the evaluation findings documented in this report. Ultimately, the HITEC Evaluation Panel reviewed and approved all of the evaluation findings.

This publication is one of 14 reports concerning the HITEC evaluation of seismic energy and dissipation devices:

- HITEC 96-02 (#40162), *Guidelines for the Testing of Seismic Isolation and Energy Dissipating Devices,* fully describes the scope and details of the HITEC evaluation program. This report is available from the American Society of Civil Engineers at 800-548-2723, 703-295-6300, or pubs@asce.org.

- This report, *Evaluation Findings for Enidine, Inc. Viscous Damper,* provides the results of the testing called for in the HITEC Evaluation Plan and measures the performance of the energy dissipator against the criteria in the *Guidelines* report.

- Also available are 10 additional reports that provide individual results for each technology tested under the seismic evaluation process.

- A *Summary of Evaluation Findings for the Testing of Seismic Isolation and Energy Dissipating Devices* report synthesizes the performance of all units submitted for evaluation by the manufacturers that participated in the HITEC evaluation and provides basic knowledge of seismic isolation and energy dissipation.

- A *Test System Overview Report* describes the test system and methodologies used to obtain the isolator characterization data. This report describes the test equipment, test procedures, instrumentation, and data processing techniques. It also provides detailed background information of interest to engineers and technicians.

Data obtained during this evaluation are available from HITEC upon request.

CHAPTER 1

Introduction

T he evaluation program for seismic isolation and energy dissipating devices was developed and executed to provide the bridge community with information and data on the dynamic performance and quality of full-scale seismic isolation and energy dissipating devices designed for highway bridge applications.

Although the concept of energy dissipation and the testing of energy dissipation devices to characterize fundamental properties is not new, the **dynamic** testing of **full-scale** devices on a national scale has not been conducted until now.

Advances in seismic isolation technology and the potential benefits seismic isolation offers have generated much interest in recent years. However, the lack of independent, full-scale dynamic characterization data and the correlation with scaled data have kept many public agencies from benefiting from what appeared to be promising technology.

In response to the heightened interest within the bridge community and the potential benefits seismic isolation offers, FHWA, Caltrans, and HITEC developed a national evaluation program for seismic isolation and energy dissipating devices.

The objectives of the evaluation program were to:

1) Implement a program of full-scale dynamic testing sufficient to characterize the fundamental properties and performance characteristics of the devices evaluated;
2) Provide guidance on the selection, use, and design of seismic isolation and energy dissipating devices for different levels of performance; and
3) Help with the development of suggested guide specifications for the use of seismic isolation and energy dissipating devices in new bridges and retrofit projects.

The evaluation program examined characteristics such as:

- Range
- Capacity
- Resilience
- Performance under service and dynamic loads
- Energy dissipation

- Functionality in extreme environments
- Resistance to accelerated aging
- Predictability of response
- Fatigue and wear
- Size effects

These properties provide the bridge designer with critical information on the suitability of the devices for specific design applications and also provide insight into the reliability, longevity, and predictability of response. Furthermore, the program addressed the ability of the vendor or manufacturer to provide a quality product and predict product response.

Acceptable ranges or limits were not established for isolator or energy dissipator test performance since these requirements are typically specified on a project-by-project basis. However, target values were defined for selected parameters so that the manufacturer could provide the appropriate component for testing.

This Evaluation Findings Report summarizes and presents data collected during the HITEC Isolator and Energy Dissipator Characterization Program (HITEC Evaluation Plan) on the five dampers submitted by Enidine, Inc. The report describes the performance characteristics of the units that were evaluated.

This report is one of 14 reports produced in conjunction with this evaluation. Please refer to the Preface for a listing of available reports.

CHAPTER 2

Test Article Description

2.1 Description of Device

The Enidine energy dissipator is a telescoping piston/cylinder device that utilizes fluid flow through orifices to absorb energy. A silicon fluid is stored in two chambers separated by the piston head. Orifices are situated in the piston head, which allow the silicon to move back and forth between the two chambers. A fluid reservoir is located in one end to control the internal pressure and provide an additional silicon fluid source in the event of a leak during dynamic motion. The force generated by this device is a result of the pressure differential across the piston head and the fluid compressibility. The unit is not designed to carry compressive or dead weight loads; it is only intended to dissipate energy caused by the relative movement of the bridge and supporting structure.

Table 2.1 summarizes the Enidine design parameters and physical characteristics of the five devices submitted for testing.

Table 2.1 Test Article Physical Properties

Test Article ID	Design Rating	Design Velocity (DV)	Design Disp. (DD)	Movement Rating (MR)	Weight	Mid-Stroke Length	Stroke Over-Travel
	(kips)	(in-sec)	(in)	(in)	(lb)	(in)	(in)
TA #1	50	20.0	6.0	2.0 (+/- 1.0)	780	66.38	0.25
TA #2	150	20.0	9.0	3.0 (+/- 1.5)	1,300	96.94	0.25
TA #3	150	20.0	9.0	3.0 (+/- 1.5)	1,300	96.94	0.25
TA #4	150	20.0	9.0	3.0 (+/- 1.5)	1,300	96.94	0.25
TA #5	240	20.0	12.0	4.0 (+/- 2.0)	2,240	123.0	0.25

2.2 Manufacturer's Data Package

The Evaluation Panel requested that each manufacturer submit a data package containing test results, working drawings, and any other data deemed necessary to assess the suitability of the manufacturing process. Manufacturers were required to provide this documentation prior to evaluation. Submitted information included:

■ WORKING DRAWINGS representing the test article (TA) and the device that the manufacturer intended to market showing all manufacturing details, dimensions, and allowable tolerances.

■ MATERIAL SPECIFICATIONS AND DATA for all components/parts with accompanying American Society for Testing and Materials (ASTM) specifications necessary to identify the test articles for future applications. (Companies that supplied test articles manufactured to foreign standards were required to submit the closest ASTM equivalent specifications.)

■ MATERIAL CERTIFICATIONS confirming that the materials used in manufacturing the test articles adhere to the material specifications.

■ PERFORMANCE PREDICTIONS and supporting design calculations.

■ ENVIRONMENTAL TEST DATA concerning each device's resistance to aging, ultraviolet light, ozone, salt spray, moisture, sand, and dust.

■ CERTIFICATE OF COMPLIANCE confirming that the test article conforms to the submitted working drawings, material specifications, allowable manufacturing tolerances, and the quality control plan. (The certificate should be supported by copies of test results performed on the devices and materials test reports for the component materials.)

■ NAME(S) AND LOCATION(S) OF THE MANUFAC-TURER.

■ QUALITY CONTROL PROGRAM SUMMARY describing the quality control procedures employed during the manufacturing process.

Table 2.2 summarizes the actual data package received from Enidine. Manufacturers were required to review ETEC's test matrix prior to testing.

Table 2.2 Enidine Submittal Summary

Description	Submitted Prior to Testing (Y/N)*	Submitted After Testing (Y/N)*	Comments
Working Drawings:			
Layout	N	N	Envelope dimensions provided 2/27/96
Assembly	N	N	
Details	N	N	
Allowable Tolerances	N	N	
Parts List	N	N	
Adapters	N	N	Envelope dimensions provided 2/27/96
Material Specifications	Y	-	
Materials Certifications	Y	-	
Performance Predictions:			
Response Equations	Y	-	In product info with HITEC application
Design Calculations	N	N	
Pretest Predictions:			
Stiffness	N	N	
Damping	N	N	
Hysteresis	N	N	
Test Data	N	N	
Test Reports	N	N	Referenced in HITEC application
Environmental Data	Y	-	Info on friction material durability in HITEC application
Certificates of Compliance	Y	-	
Name and Location of Manufacturer	Y	-	
Q/C Program Summary	Y	-	

* Y=Yes; N=No

CHAPTER 3
Test Setup and Test Plan

3.1 Installation

Adapters

Enidine provided sketches of the adapters and spacers. The drawings conformed to all HITEC/ETEC requirements. The test articles were fabricated and shipped by Enidine.

Handling Requirements

The manufacturer did not specify special handling requirements for the test articles.

Installation Requirements

Enidine did not specify any special installation requirements.

Installation Problems

The adapters and test articles were installed without incident.

3.2 Performance Testing

Testing of the Enidine energy dissipators was conducted in accordance with the requirements set forth in the HITEC Evaluation Plan, which consists of nine separate tests. Each test provided a means for evaluating a specific characteristic of the test item(s):

Test 1 – Performance Benchmark: To verify experimentally the initial response characteristics, damping, and number of loading cycles required to stabilize response.

Test 2 – Compressive Load Dependent Characterization: To quantify experimentally the effects of varying compressive loads on the performance characteristics, specifically stiffness, damping, and energy dissipation per cycle (EDC). This test is only applicable to isolators; therefore, it was not used in the evaluation of this energy dissipator.

Test 3 – Frequency Dependent Characterization: To determine experimentally dynamic performance characteristics at varying frequencies in the primary direction of operation. This test is only applicable to isolators; therefore, it was not used in the evaluation of this energy dissipator.

Test 4 – Frequency Dependent Characteristics: To determine dynamic performance characteristics and verify the constitutive laws.

Test 5 – Fatigue and Wear: To evaluate the potential seismic performance changes resulting from 10,000 cycles of service movements (temperature and live load fluctuations).

Test 6 – Environmental Aging: To verify experimentally seismic performance after exposing the device to a salt spray environment.

Test 7 – Dynamic Performance Characteristics at Temperature Extremes: To assess the effects of extreme temperature on the performance characteristics.

Test 8 – Durability: To assess component durability resulting from a moderate number of strong motion cycles.

Test 9 – Ultimate Performance: To determine experimentally ultimate displacement and margins of safety.

ETEC performed the applicable tests on each of the test articles submitted by Enidine. The test matrix specified by the HITEC Evaluation Plan for damping devices is summarized in Table 3.1. Test details are described in the following sections.

Table 3.1 HITEC Program Energy Dissipator Test Matrix

TA #1 (50 kip)	Test 1	Test 4	Test 8	Test 9	N/A	N/A
TA #2 (150 kip)	Test 1	Test 4	Test 7	Test 8	Test 9	N/A
TA #3 (150 kip)	Test 1	Test 4	Test 7*	Test 8	Test 9	N/A
TA #4 (150 kip)	Test 1	Test 4	Test 5	Test 6	Test 4	Test 9
TA #5 (240 kip)	Test 1	Test 4	Test 8	Test 9	N/A	N/A

Note: Test numbers do not correspond to the testing sequence. Table 3.1 shows the actual test sequence, left to right, for each test article submitted.

* Test 7 (for hot temperatures) was to be performed on TA #3 only if it was fabricated from different materials than TA #2.

CHAPTER 4

Data Summary, Analysis, and Review

This chapter summarizes results for each test performed on the Enidine energy dissipators. Test results and specific details regarding the individual component tests are discussed. Idealized damping and velocity coefficients, based on the viscous damping, were calculated using the methodology shown in Appendix B.

Results from each test are reported in the following order:

1) Purpose
2) Test Procedure
3) Test Variances
4) Data Summary
5) Test Observations

Test results and normalized test results were plotted for Tests 1, 4, and 8. Normalized plots were used to directly compare dampers of varying sizes. Dynamic performance characteristics such as force degradation and EDC for various size dissipators when normalized can show performance trends and variations. Variations in performance may be attributed to manufacturing processes, material variations, size effects, or other factors.

4.1 Test 1 – Performance Benchmark

Purpose

To experimentally verify the initial performance characteristics, damping, and number of loading cycles required to stabilize response. This number of cycles is referred to as the number of shake-down cycles.

Procedure

ETEC applied 10 fully reversed cycles at the design displacement (DD) at a frequency corresponding to a 2.0 second period. If the device did not stabilize, the manufacturer was to be notified that no further testing would be undertaken.

$$F_i = \text{Peak lateral load for the } i^{th} \text{ cycle}$$

The factor F_2/F_{10} was used to indicate whether shake-down occurs. Currently, there are no definite limits to determine whether shake-down has occurred. However, based on a consensus of the Technical Evaluation Panel for this testing program, it was established that shake-down occurs if $0.7 < F_2/F_{10} < 1.3$, as documented in the HITEC Evaluation Plan. The Panel later decided that F_3 is more appropriate than F_2. Therefore, Table 4.1 uses the third cycle rather than the second cycle. The table also presents the values of EDC_3/EDC_1 and EDC_{10}/EDC_1 for informational purposes.

The resulting data were used to determine shake-down characteristics of the test article and confirm that minimum damping requirements were satisfied for the devices or combination of devices evaluated.

Test Variances

During this test, variances from the standard test procedure were necessary. The variances are described as follows:

Variance #1: Two half-cycle tests were performed prior to the start of Test 1.

Reason: The Enidine dissipators have small displacement tolerances. Therefore, it was important to set the device to the exact mid-stroke point prior to testing to avoid accidentally exceeding the stroke limits during testing.

Impact: None.

Variance #2: The Performance Benchmark test was not used to terminate the evaluation if a test article did not meet the evaluation protocol shake-down criteria.

Reason: The HITEC Panel deemed the requirements as described above and in the Evaluation Plan to be too stringent.

Impact: Tests 2 through 9 were executed. Additional information on damper performance was obtained.

Data Summary

The results from Test 1, which was performed to characterize test article response stability, are summarized in Table 4.1 and Figure 4.1 (a and b).

Test Observations

Both Test Article #1 and #5 failed during the 10th cycle of Performance Benchmark Test when a seal burst causing hydraulic fluid to spray from the devices. Although the units failed during this test, the performance did not appear to degrade, i.e., the force-deflection or force-time plots did not exhibit anomalies prior to failure. Test Article #1 was stable after four cycles whereas the other four energy dissipators did not shake down, see Figure 4.1. Performance of the three identical energy dissipators did not agree with each other.

According to Enidine, the dampers failed because a small relief orifice had been inadvertently omitted. The dampers are designed to absorb energy by converting work energy into heat. As the testing progressed, the temperature of the internal fluid increased, causing the fluid to expand. Without the relief orifice, the internal pressure increased until a tie rod stretched, allowing the cylinder head, and thus O-ring seal, to move out of place. Once the tie rods stretched, the internal fluid leaked out under high pressure.

4.2 Test 4 – Frequency Dependent Characterization

Purpose

To determine dynamic performance characteristics and verify the constitutive laws.

Procedure

Three fully reversed cycles of the maximum displacement were applied at each of the following harmonic frequencies: 0.05

Table 4.1 Test 1 — Performance Benchmark

Test Article ID	F_3/F_1	F_{10}/F_3	EDC_{10}/EDC_1	EDC_3/EDC_1
TA #1 (50 kip)	0.82	0.89	0.76	0.89
TA #2 (150 kip)	0.77	0.66	0.53	0.80
TA #3 (150 kip)	0.81	0.66	0.48	0.83
TA #4 (150 kip)	0.90	0.65	0.64	0.93
TA #5 (240 kip)	0.93	0.69	0.63	0.91

F_i = Peak force during i^{th} cycle.

EDC_i = Energy dissipation during i^{th} cycle.

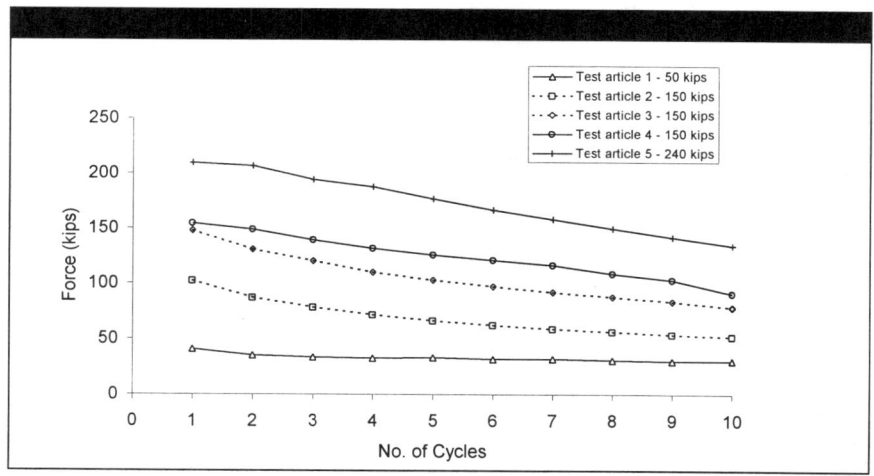

Figure 4.1(a) Force Stability During Performance Benchmark Test: Force Degradation

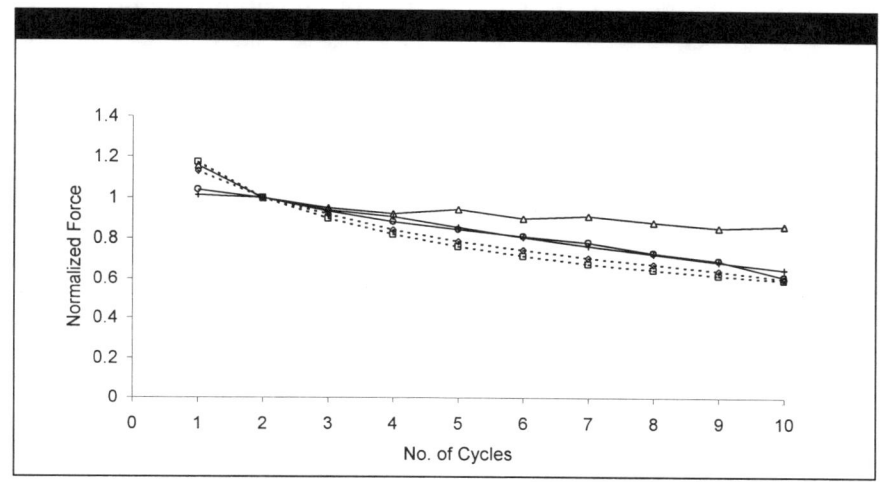

Figure 4.1(b) Force Stability During Performance Benchmark Test: Normalized Force Degradation

Hz, 0.2 Hz, 0.5 Hz, 1.0 Hz, and 2.0 Hz (corresponding to 20.0 second, 5.0 second, 2.0 second, 1.0 second, and 0.5 second periods). The displacement ranges during the test may have been limited to ensure the maximum velocity of the device or test equipment was not exceeded.

The data were used to quantify the performance of energy dissipating devices at various frequencies and velocities and confirm analytical predictions of the form $F = CV^n$, where F is the force, C is the Damping Coefficient, V is the velocity, and n is the Velocity Coefficient. The value of C for a given damper is dependent upon the units of velocity (e.g. British units vs. Metric). The value of n is a function of the damper design, which can be achieved through manufacturing techniques and is usually less than 1.0.

Test Variances

During this test, variances from the standard test procedure were necessary. The variances are described as follows:

Variance #1: The 0.5, 1.0, and 2.0 second period tests for Test Articles #2, #3, and #4 were performed at reduced displacement so the Design Velocity (DV) would not be exceeded.

Reason: To perform the tests at the design displacement and required frequency would have required a velocity that exceeded the DV. The Evaluation Plan stipulated that the DV shall not be exceeded during this test.

Resolution: These tests were performed at the desired frequency, but the displacement was reduced such that the test velocity was equal to or less than DV.

Impact: Direct performance comparisons for the full frequency range are not possible.

Variance #2: During the first 2.0 second period test for TA #4 not all the required cycles were completed.

Reason: Operator error. Limited amount of hydraulic fluid left in the accumulator.

Resolution: The accumulator was recharged the test repeated.

Impact: None.

Data Summary

The results of Test 4, which was performed to characterize velocity and damping coefficients and the EDC as a function of forcing frequency, are summarized in Table 4.2 and Figure 4.2. The data for Figure 4.2 were taken from the first quarter cycle of the 2.0 second period test.

Test Observations

Although all three dampers tested for Frequency Dependency Characterization were the same size units, they all displayed different performance characteristics. Test Article #2 had a velocity coefficient less than one, whereas for Test Article #3 it was generally greater than one. The data from Test Article #4 displayed an inconsistent velocity coefficient. In addition, none of the three 150 kip units achieved the target force capacity.

Table 4.2 Test 4 – Frequency Dependent Characterization (Second Cycle)

Test Article ID	Period (sec)	Displacement (in)	Velocity (in/sec)	Velocity Coeff — n	Damp Coeff — C	EDC (in-kips)
TA #1 (50 kip)	20.0	6.0	1.9			
	5.0	6.0	7.5			
	2.0	6.0	18.8		*	
	1.0	3.18	20.0			
	0.5	1.59	20.0			
TA #2 (150 kip)	20.0	9.0	2.8	0.86	7.5	533.1
	5.0	9.0	11.3	0.76	8.0	1486
	2.0	6.37	20.0	0.76	7.5	1515
	1.0	3.18	20.0	0.82	6.3	704.5
	0.5	1.59	20.0	0.78	7.1	370.7
TA #3 (150 kip)	20.0	9.0	2.8	1.53	5.2	652.7
	5.0	9.0	11.3	1.27	3.2	1836
	2.0	6.37	20.0	1.17	2.8	1740
	1.0	3.18	20.0	0.94	5.8	913.0
	0.5	1.59	20.0	0.88	6.7	459.0
TA #4 (150 kip)	20.0	9.0	2.8	0.77	17.9	118.7
	5.0	9.0	11.3	0.62	19.5	2707
	2.0	6.37	20.0	0.75	12.9	2506
	1.0	3.18	20.0	1.04	5.5	1137
	0.5	1.59	20.0	1.96	0.3	484.8
TA #5 (240 kip)	20.0	12.0	3.77			
	5.0	12.0	15.1			
	2.0	6.37	20.0		*	
	1.0	3.18	20.0			
	0.5	1.59	20.0			

* Test 4 was not performed on TA #1 and TA #5 due to earlier failure.

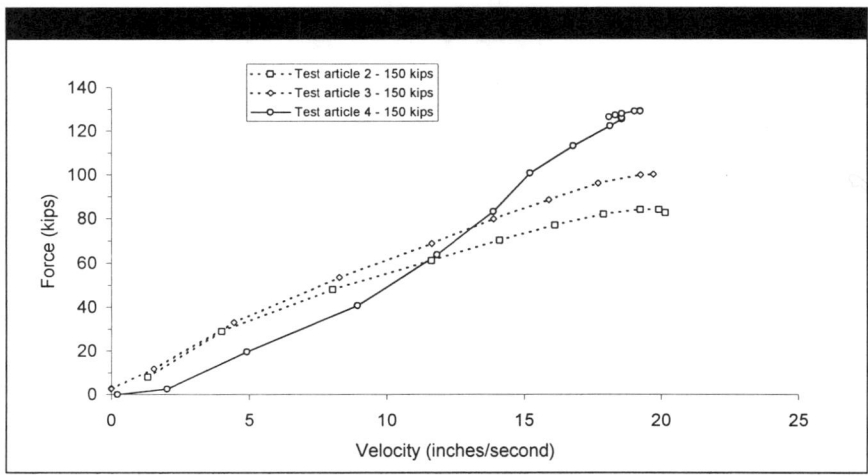

Figure 4.2 Frequency Test Data: Force vs. Velocity

4.3 Test 5 – Fatigue and Wear

Purpose

To evaluate the potential seismic performance changes resulting from 10,000 cycles of service movements (temperature and live load fluctuations).

Procedure

The device was subjected to a minimum of 10,000 cycles of simulated displacement representing the movement rating (MR) of the device specified by the manufacturer and applied at a frequency corresponding to a 10.0 second period or other frequency deemed acceptable to the manufacturer. The Evaluation Plan required that the speed of application be at least 4.5 inches per minute. The resulting data were used to quantify damage or degradation arising from the application of high-frequency, small-displacement loading.

Test Variances

No variances were reported.

Data Summary

Deterioration from fatigue and wear was not evident based upon visual inspection.

Test Observations

No specific damage was observed. The test article was sent to an outside laboratory for environmental aging (Test 6).

Table 4.3 Tests 5 and 6 – Fatigue and Wear and Environmental Aging (Second Cycle)

Period (Seconds)	Velocity Coeff — n		Damping Coeff —C		EDC (in-kips)	
	Before**	After	Before**	After	Before**	After
20.0	0.77	0.86	17.9	20.9	1187	1494
5.0	0.62	0.69	19.5	19.2	2707	3056
2.0*	0.75	0.77	12.9	13.0	2506	2659
1.0*	1.04	0.93	5.5	7.4	1137	1116
0.5*	1.96	1.88	0.3	0.5	484.8	541.0

* Performed at reduced displacement.

** "Before" data are obtained from Table 4.2.

4.4 Test 6 – Environmental Aging

Purpose

To verify experimentally seismic performance after exposure to a salt spray environment.

Procedure

The device was exposed in a salt spray chamber for 1,000 hours in accordance with the requirements of ASTM B117. The accelerated aging testing examined the degradation of the device resulting from key environmental factors.

Test Variances

No variances were reported.

Data Summary

An outside laboratory performed salt spray conditioning on TA #4 (150 kip unit). The device was returned for testing without the optional warm water rinsing. Test 4, Frequency Dependent Characterization, was then used to assess changes in performance.

The accelerated aging effects of fatigue and wear (Test 5) and salt spray (Test 6) on velocity coefficients, damping coefficients, and EDC are summarized in Table 4.3.

Test Observations

Following the combined 10,000 cycle and salt spray test the damping coefficients and EDC generally increased, while the velocity coefficients did not exhibit a pattern of change. Overall, the changes in performance were not significant.

4.5 Test 7 – Dynamic Performance Characteristics at Temperature Extremes

Purpose

To assess the effects of extreme temperature on the performance characteristics.

Procedure

Three fully reversed cycles of the design displacement were applied at a frequency corresponding to a 2.0 second period at the upper and lower temperature extremes specified by the manufacturer. The temperature range of interest for this evaluation program was from -40° F to 120° F. However, if specific devices could not perform at these temperature extremes, manufacturer-specified temperature limits were to be used.

Further test procedures for handling hot or cold test articles were subsequently developed as follows:

1) The test article and mounting hardware were placed in the heating/cooling unit for the prescribed number of hours;
2) The test article and mounting hardware were installed in the test rig within 75 minutes of being removed from the thermal chamber; and
3) Testing was performed within five minutes after installation was complete.

Table 4.4 Test 7 – Dynamic Performance Characteristics at Temperature Extremes (Second Cycle)

Performance Parameters	Cold Temperature 24 hrs @ -40° F (TA #2)	Ambient Temperature 70° F (TA #2)/(TA #3)*	Hot Temperature 24 hrs @ 120° F (TA #3)
Velocity Coeff — n	0.60	0.76/1.17	1.09
Damping Coeff — C	22.3	7.5/2.8	4.1
EDC (in-kips)	2930	1515/1740	2079

* Data obtained from Table 4.2 for 2.0 second period.

Table 4.5 Test 8 – Durability

Test Article	F_1(kips)	F_5(kips)	F_{10}(kips)	F_{15}(kips)	F_{20}(kips)
TA #1 (50 kip)	**	**	**	**	**
TA #2 (150 kip)	101	66	37	29	**
TA #3 (150 kip)	80	65	58	51	47
TA #5 (240 kip)	**	**	**	**	**
** These cycles were not completed due to an earlier failure.					

The resulting data quantified changes in performance that occurred as a result of changes in ambient temperature conditions.

Test Variances

No variances were reported.

Data Summary

The manufacturer rated the temperature operation range for TA #2 from –40 °F to 40 °F and for TA #3 from 30 °F to 120 °F. Cold temperature testing was performed on TA #2 and hot temperature testing was performed on TA #3. The effects of temperature on velocity and damping coefficients and EDC are summarized in Table 4.4.

Test Observations

Damping and EDC increased during the cold temperature test and the hot temperature test. The velocity coefficient decreased during the cold temperature test and remained relatively constant during the hot temperature test.

4.6 Test 8 – Durability

Purpose

To assess component durability resulting from a moderate number of strong motion cycles.

Procedure

Twenty fully reversed cycles at 100 percent maximum design displacement were applied at a frequency corresponding to a 2.0 second period. The resulting data were used to assess performance degradation, quantify useable margins, and provide insight into the suitability of the device for applications where a large number of high-level aftershocks can be expected.

Test Variances

During this test, variances from the standard test procedure were necessary. The variances are described as follows:

Variance #1: Two incomplete tests were run on TA #3. Both tests were stopped at the end of the 12th cycle.
Reason: There was a problem with the computer storage buffer.
Resolution: The successful completion of Test 8 was postponed and the test article was subjected to Test 9.
Impact: A complete data set with 20 continuous cycles was not obtained at this point.

Variance #2: Following the completion of Test 9, TA #3 was subjected to Test 8.
Reason: To obtain data on 20 continuous cycles for Test 8.
Impact: The appropriate data were collected. Although the test article experienced 30 additional cycles before the successful completion of Test 8, the impact to its overall performance was expected to be minimal.

Data Summary

In addition to demonstrating the ability to withstand a significant number of design displacement loadings, Test 8 can also be viewed as an extension of the stability test (Test 1). The results are summarized in Table 4.5 and Figure 4.3(a-b).

Test Observations

Test Article #2 failed when a seal ruptured during the 16th cycle of the test. Misting of the hydraulic fluid was first observed after the 9th test cycle.

Due to computer problems, Test Article #3 was tested three times to successfully complete 20 continuous cycles for Test 8. Following the first two incomplete tests, the test article was subjected to Test 9. Subsequently, Test 8 was completed successfully. During the first incomplete test, the test article exhibited much higher force capacities and degradation than the next two Test 8 runs.

Table 4.6   Test 9 – Ultimate Performance

Test Article ID	DV (in/sec)	Max. Velocity Achieved (in/sec)	Actuator Force (kips) at Selected Velocities (in/sec)			
			0.5 x DV Actuator Force (Velocity)	1.0 x DV Actuator Force (Velocity)	1.5 x DV Actuator Force (Velocity)	2.0 x DV Actuator Force (Velocity)
(#1) 50 kip	20	**	**			
(#2) 150 kip	20	**	**			
(#3) 150 kip	20	37	45 (10)	69 (20)	103 (30)	115* (40)
(#4) 150 kip	20	34.6	45 (10)	145 (20)	176 (30)	200* (40)
(#5) 240 kip	20	**	**			

* Testing machine did not achieve target velocity. The actuator force was determined by extrapolation from test data.

** Not tested due to an earlier failure.

4.7 Test 9 – Ultimate Performance

Purpose

To determine performance characteristics when test articles are subjected to loads over design values.

Procedure

During this test the device was loaded at a frequency corresponding to a 2.0 second period with increasing displacement amplitude until "failure" occurred. Each cycle increased the displacement amplitude by 1.1 times the preceding cycle, e.g., $(1.1)^0$x DD, $(1.1)^1$ x DD, $(1.1)^2$x DD, $(1.1)^3$x DD ... $(1.1)^n$x DD for the nth cycle. For some devices, physical failure may be an actual "failure," whereas for other devices it was a programmed physical stop or ultimate restraint beyond which the device could not operate.

The data were used to characterize performance and provide insight into the stability of the device under high loading states. Failure of a hydraulic energy dissipator was defined to occur when 1) a major hydraulic leak is observed, 2) the piston assembly hits a physical stop, or 3) a significant change in performance is observed. Velocity and damping coefficients were determined throughout the entire range.

Test Variances

During this test, variances from the standard test procedure were necessary. The variances are described as follows:

Variance: The test procedure for piston/cylinder type energy dissipators was redefined to create velocities within the useable stroke range of the unit that went to 2 x DV.

Reason: The original test procedure was deemed inappropriate for these types of dissipators, since their pistons will not permit displacements greater than the design displacement.

Resolution: The following test methodology was defined to create velocities up to 2 x DV within the useable stroke range of the test articles:

The actuator was retracted to the compressive limit of the test article (-DD) very slowly and then accelerated to its fully extended position (+DD) at a rate that will produce the target velocity at a point very near the fully extended position. The motion is then immedi-

Table 4.7 Comparison of Predicted and Measured Performance

Test Article Size	Velocity Coeff —n		Damping Coeff —C		EDC (in-kips)	
	Design	Actual*	Design	Actual*	Design	Actual*
150 kip	0.50	0.76	33.54	7.5	**	1515

* "Actual" data were taken from 2.0 second period data for TA #2 in Table 4.2.

** Not provided.

ately halted at this position (+DD). The actuator was then accelerated to its fully retracted position (-DD) at the same rate, producing the target velocity in the other direction. The motion was immediately halted at this position. The target velocity was incremented from 0.5 x DV up to a maximum of 2.0 x DV. One complete cycle was executed in this manner for each target velocity.

Impact: The ultimate performance of the units was tested using velocities that exceed the design values instead of displacements.

Data Summary

Data obtained during this test are summarized in Table 4.6.

Test Observations

Two test articles were subjected to loadings that required velocities substantially above their design velocities. Both test articles completed the testing without visible signs of distress. Although the test articles were of the same size, their performance was markedly different.

4.8 Test Predictions

Prior to testing, Enidine predicted velocity coefficients, damping coefficients, and EDC for their devices. These performance characteristics were measured for the 150 kip units during the HITEC testing. A comparison of the predicted vs. measured values is provided in Table 4.7.

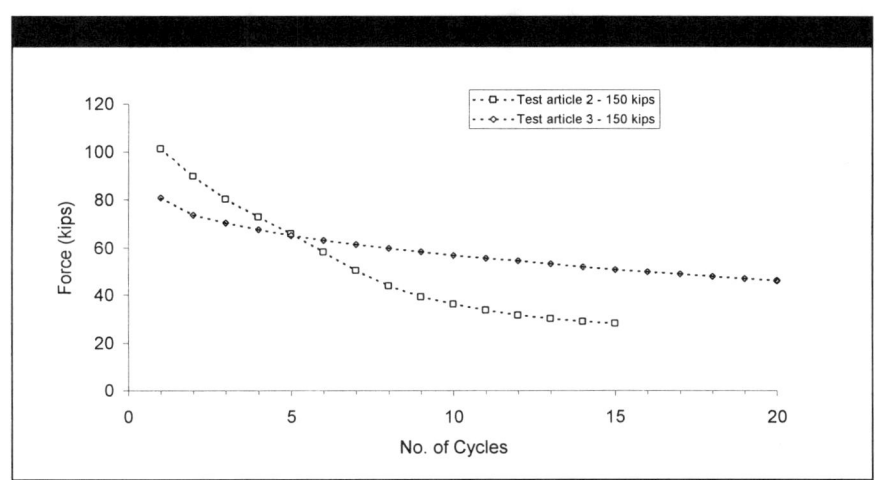

Figure 4.3(a) Force Degradation During Durability Test: Force Degradation

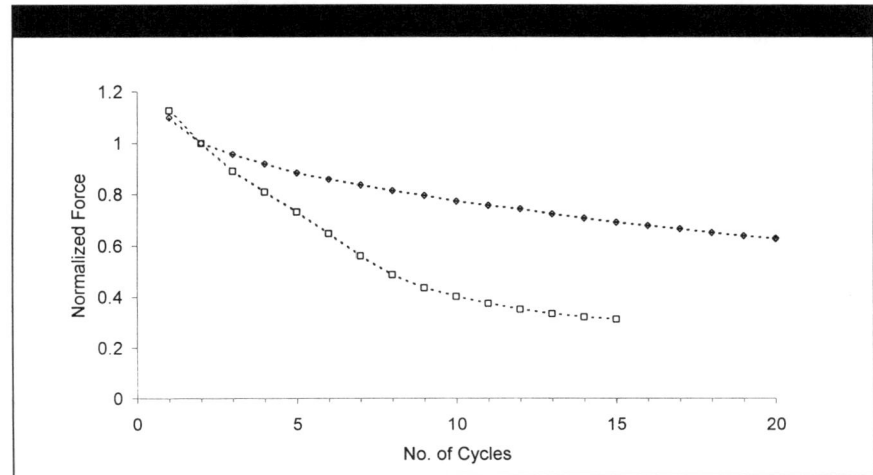

Figure 4.3(b) Force Degradation During Durability Test: Normalized Force Degradations

CHAPTER 5

Summary

In March 1996, HITEC published *Guidelines for the Testing of Seismic Isolation and Energy Dissipating Devices*, an Evaluation Plan, under the guidance of a Technical Evaluation Panel. This Plan described test methods used to evaluate the performance of several different types of isolators and dampers manufactured by a number of companies. The Evaluation Plan specified design parameters for the isolators and dampers, tests to be performed, and testing instructions. Full-scale, dynamic tests were specified to measure performance benchmarks, compressive load dependency, frequency dependency, fatigue and wear effects, effects of environmental aging, dynamic performance at extreme temperatures, durability, and ultimate performance. Testing was subsequently completed for 11 technologies from 10 companies.

Enidine, Inc. manufactures and sells fluid viscous dampers to dissipate the energy associated with earthquakes. As part of this evaluation, Enidine manufactured five energy dissipators, which were subjected to testing at the ETEC facility. The dissipators consisted of three sizes with capacities of 50 kips, 150 kips (three dissipators), and 240 kips. The design displacement for each device was six, nine, and twelve inches, respectively.

Based on the test results, the following observations were made:

- The 50 kip and 240 kip dissipators failed during the first test, the Performance Benchmark Test. The three 150 kip devices, although of the same size, did not exhibit similar performance characteristics during this test.

- During tests performed at varying frequencies, the 150 kip devices again exhibited dissimilar performances.

- Visual observations of a unit that was exposed to 10,000 cycles of service movements did not find any damage. The same test article was then exposed to a salt spray environment. Test results showed that changes in performance were not significant. However, the damping coefficients and EDC tended to increase.

- Damping and EDC increased during both the cold and hot temperature tests. The velocity coefficient decreased for the cold temperature test, and remained relatively constant during the hot temperature test.

- The first 150 kip device failed during the 16th cycle of the durability test. The second 150 kip device completed two partial runs of this test, the ultimate performance test, and then a complete run of the durability test.

- Two 150 kip test articles were subjected to loadings that required velocities substantially above their design velocities. Both test articles completed the testing without visible signs of distress. Although the test articles were the same size, their performance was markedly different.

It should be noted that according to Enidine, an internal relief orifice was inadvertently omitted from the test articles. The orifice is designed to relieve the internal pressures that caused failure in three out of the five test articles submitted for testing. Since the HITEC testing, Enidine has enhanced their damper design. The relief orifice is now installed at assembly and confirmed during an Enidine pre-production test. The tie rod design has also been modified. At the time of this writing, Enidine is waiting for testing to be conducted on the new design. HITEC will produce a report to document the new findings.

Appendix A

Selected Test Plots

This appendix contains the characteristic plots of selected test runs. Additional plots listed in Appendix C are available upon request from HITEC.

Over 80 plots were generated for each damper tested. A select number of representative plots are provided in the body of this report for TA's #1 (50K), #3 (150K), and #5 (240K). The test numbers given on these plots are internal numbers for the testing facility and are not meaningful within the context of this report.

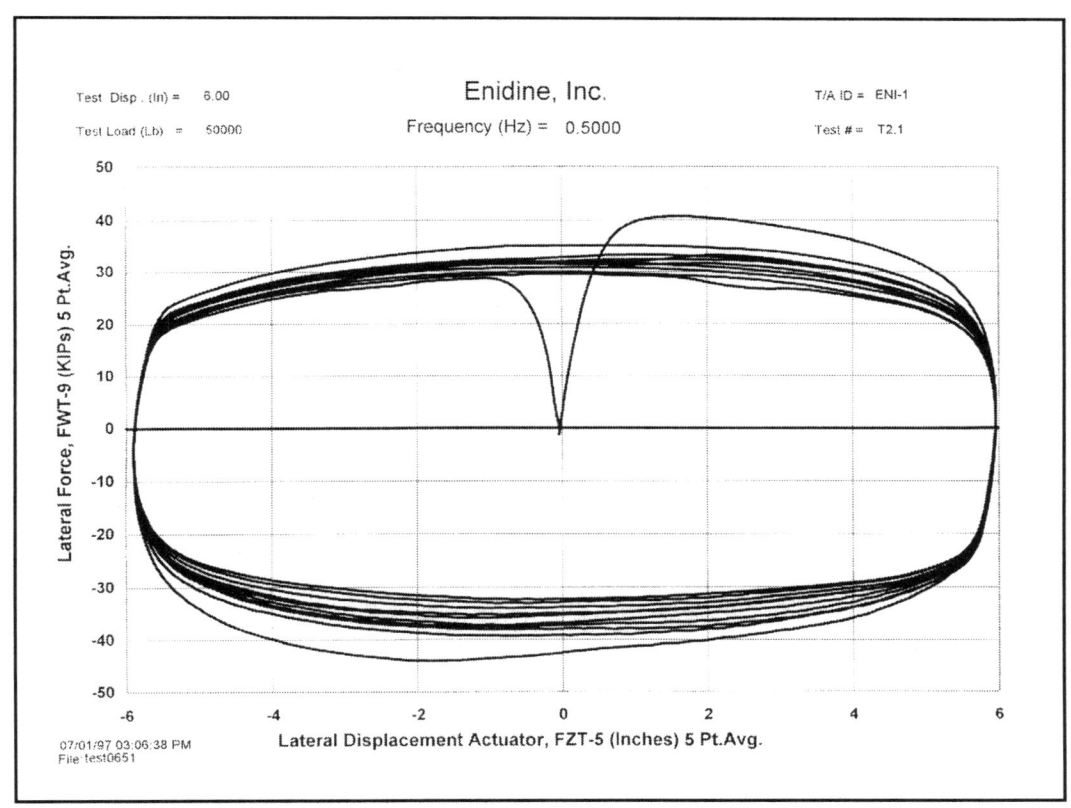

A.1   10 Cycle Stability Force Deflection of Test Article #1 (50 kip) Plot

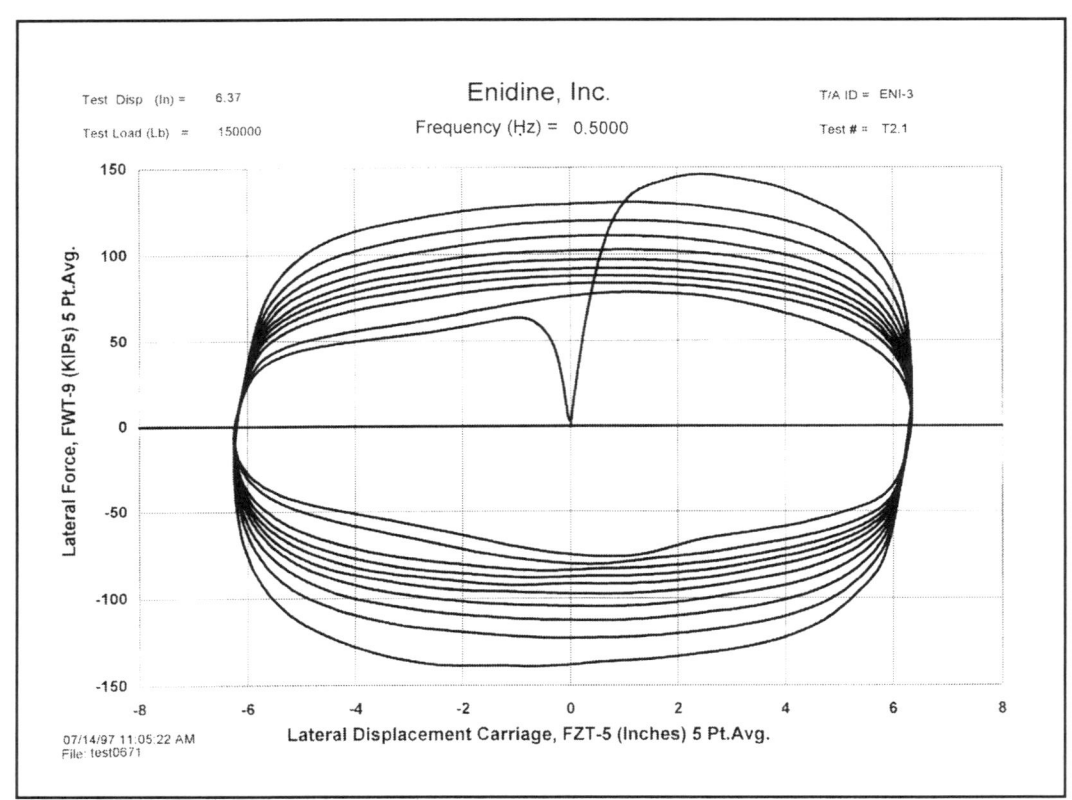

A.2   10 Cycle Stability Force Deflection of Test Article #3 (150 kip) Plot

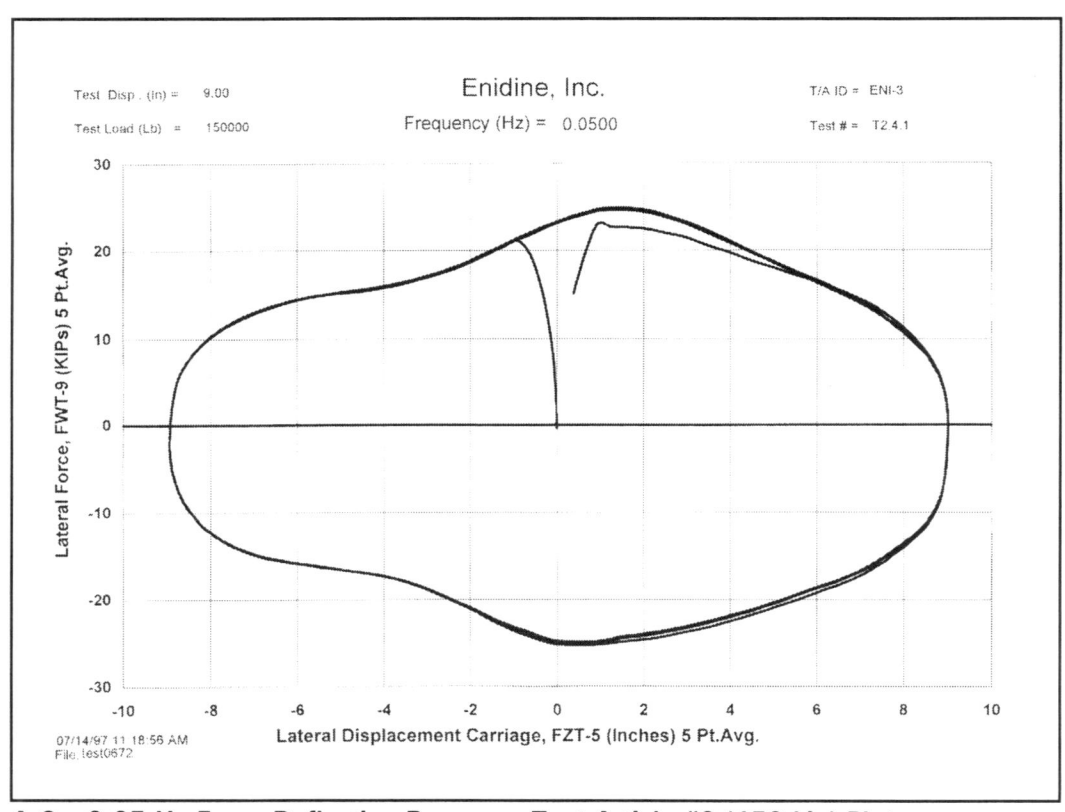

A.3   0.05 Hz Force Deflection Response Test Article #3 (150 kip) Plot

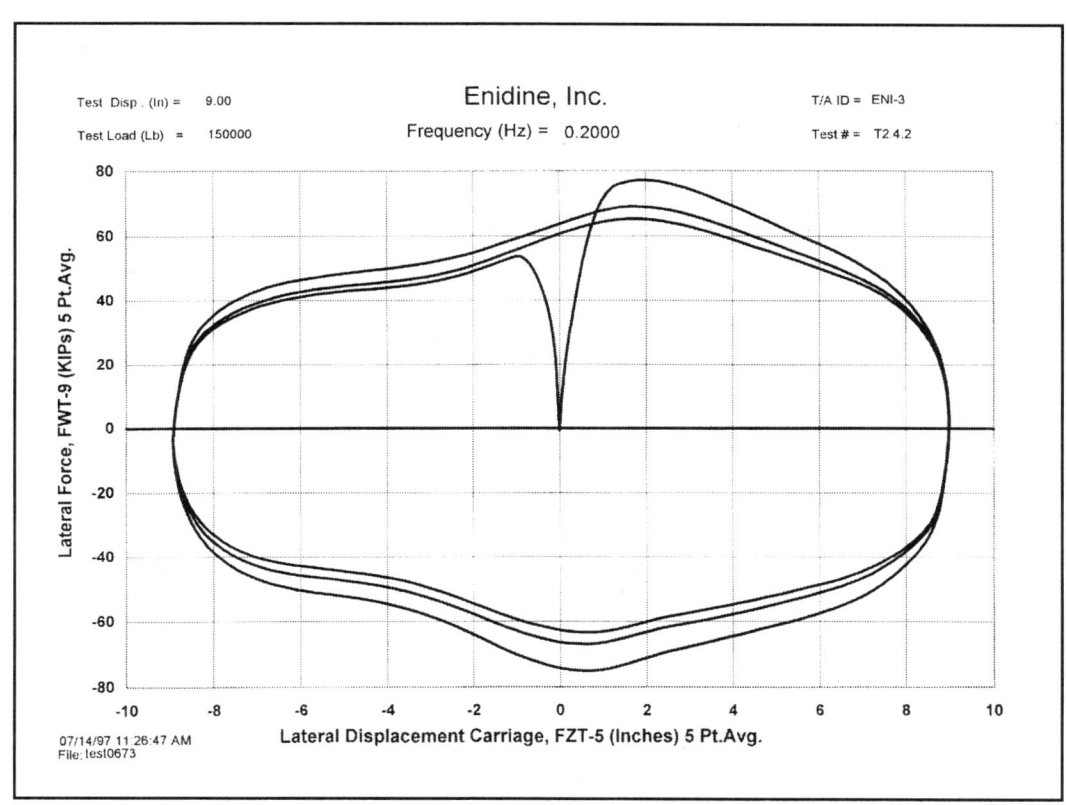

A.4   0.20 Hz Force Deflection Response Test Article #3 (150 kip) Plot

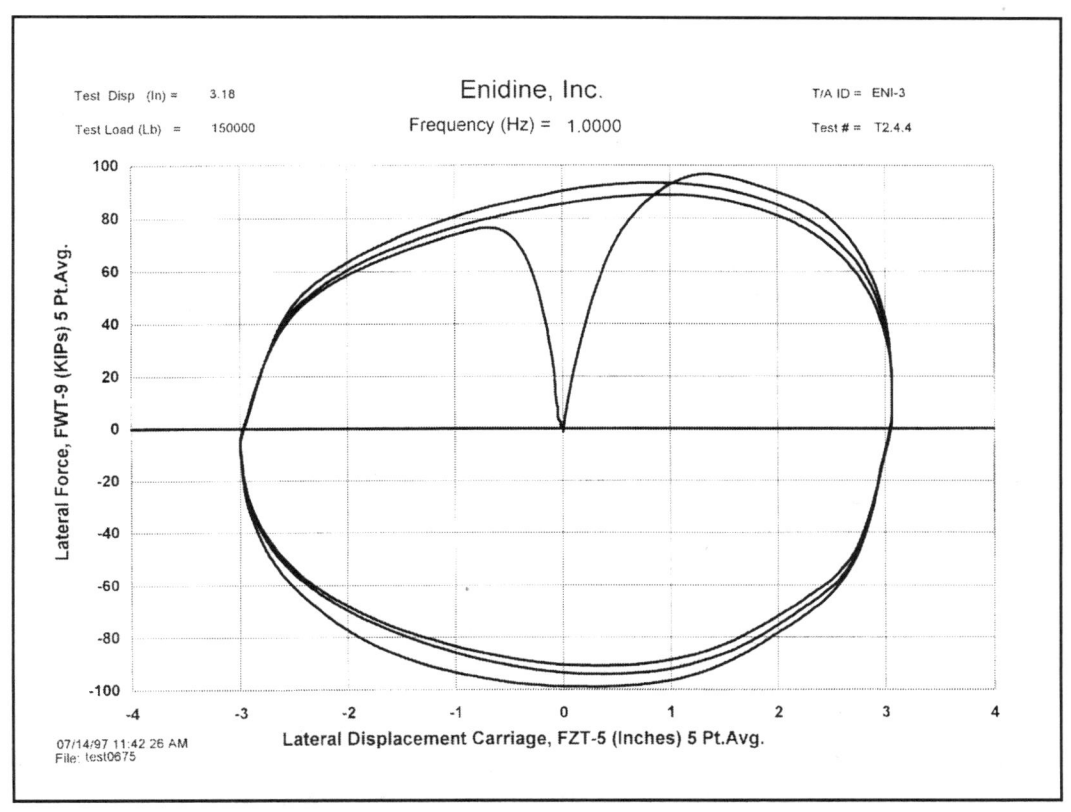

A.5   1.0 Hz Force Deflection Response Test Article #3 (150 kip) Plot

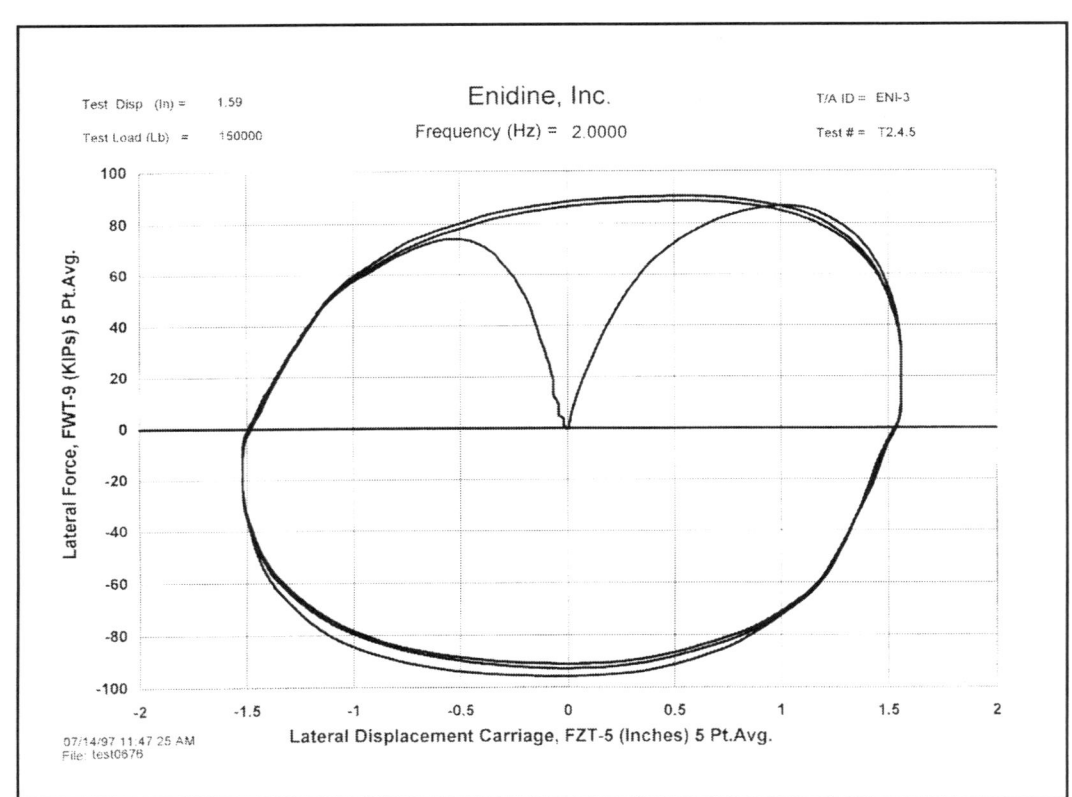

A.6   2.0 Hz Force Deflection Response Test Article #3 (150 kip) Plot

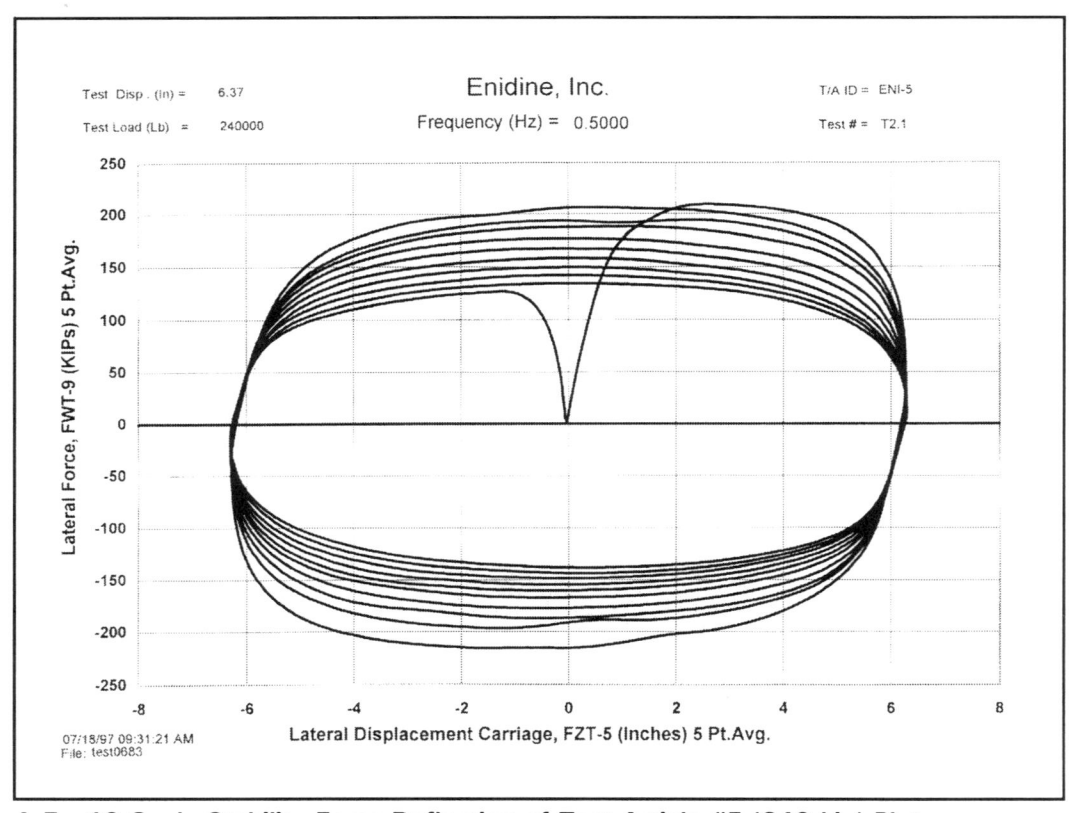

A.7   10 Cycle Stability Force Deflection of Test Article #5 (240 kip) Plot

Appendix B

Reported Damping and Velocity Coefficients

Assume:
$$F(X) = F_o X^n$$
$$X = X_o \sin(wt)$$

Therefore, for harmonic motions the area under the force-displacement function is:

$$A = \oint F(X)dX$$

$$A(n) = (F_n/w)(X_o w)^{n+1} \int_0^{2\pi} (\cos(U))^{n+1} dU$$

$$A/(4F_n X_o) = \int_0^{\pi/2} (\cos(U))^{n+1} dU$$

$$A/(4F_n X_o) = B(n)$$

Or

$$N = f(A/4 F_n X_o) = f(Q)$$

$$n = -24.68Q^3 + 71.07Q^2 - 72.35Q + 25.96$$
$$\text{Where } Q = A/4 F_n X_o$$

The above equation is used to generate the following plot:

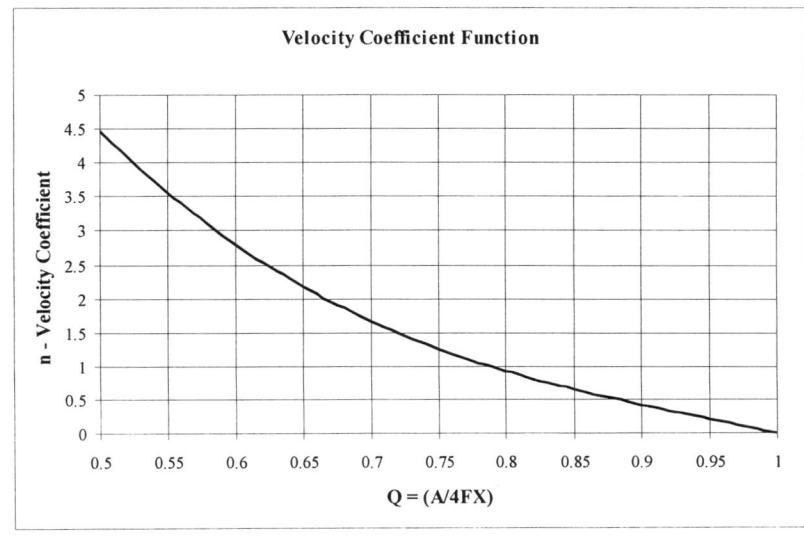

Velocity Coefficient Function

Appendix C

Available Data Plots

The following plots are available from HITEC upon request:

Test 1 - Performance Benchmark

Lateral Force vs. Lateral Displacement
Lateral Force vs. Time
Average Vertical Displacement vs. Time
Lateral Displacement vs. Time
Total Compressive Force vs. Time
Total Compressive Force vs. Lateral Displacement
Average Vertical Displacement vs. Lateral Displacement

Test 4 - Frequency Dependent Characterization

Lateral Force vs. Lateral Displacement
Lateral Force vs. Time
Lateral Displacement vs. Time
Average Vertical Displacement vs. Time
Total Compressive Force vs. Time
Total Compressive Force vs. Lateral Displacement
Average Vertical Displacement vs. Lateral Displacement

Test 5 - Fatigue and Wear

No Plots Recorded

Test 6 - Environmental Aging

No Plots Recorded

Test 7 - Dynamic Performance at Temperature Extremes

Lateral Force vs. Lateral Displacement
Lateral Force vs. Time
Lateral Displacement vs. Time
Average Vertical Displacement vs. Time
Total Compressive Force vs. Time
Total Compressive Force vs. Lateral Displacement
Average Vertical Displacement vs. Lateral Displacement

Test 8 - Durability

Lateral Force vs. Lateral Displacement
Lateral Force vs. Time
Lateral Displacement vs. Time
Average Vertical Displacement vs. Time
Total Compressive Force vs. Time
Total Compressive Force vs. Lateral Displacement
Average Vertical Displacement vs. Lateral Displacement

Test 9 - Ultimate Performance

Lateral Force vs. Velocity
Lateral Force vs. Time

Glossary

AASHTO. American Association of State Highway and Transportation Officials.

ASCE. American Society of Civil Engineers.

ASTM. American Society for Testing and Materials.

Caltrans. California Department of Transportation.

CERF. Civil Engineering Research Foundation.

Damper. See Energy Dissipator.

Damping. The ability to dissipate energy.

Design Displacement (DD). The maximum lateral displacement under seismic loading.

Design Compressive Load (DCL). The maximum design vertical load (dead load, live load, overturning, etc.).

Design Velocity (DV). Velocity at the rated load.

Equivalent Damping. Value of equivalent viscous damping corresponding to the energy dissipated during cyclic response at the design displacement of the isolator.

Energy Dissipation per Cycle (EDC). Area under force deflection hysteresis loop.

Energy Dissipator. A device that is used to dissipate energy by friction or viscous flow.

ETEC. Energy Technology Engineering Center.

FHWA. Federal Highway Administration.

HITEC. Highway Innovative Technology Evaluation Center.

Isolator. A device that is used to isolate a structure from seismic ground motion.

Motion-Controlled. Specified motion that an isolator or energy dissipator is forced to follow in a test.

Movement Rating (MR). The small displacement range (lateral) of the device due to temperature and live load fluctuations (excluding earthquakes).

Number of Shake-Down Cycles. Number of fully reversed motion-controlled cycles required for an isolator or energy dissipator to repeat its performance or stabilize its cyclic response.

TA(s). Test article(s).

TRB. Transportation Research Board.